Beekeeping:
Step-By-Step Guide For Beginner And Advanced Beekeepers

Table of content:

Introduction

We have all heard of the land that is flowing with milk and honey, and when we close our eyes, we think of how wonderful that would be. To have as much honey as you want, when you want – it's more than just a dream, it would make your life much easier.

There are many benefits to adding honey into your diet regularly, from fighting off viruses and infections to slimming down and keeping the weight off. When it comes to honey, all you can see are the good, as there is very little bad.

But, honey is expensive. You have shopped around, you have tried to find the best you can get without breaking the bank, but you are consistently coming home with a small jar and an empty wallet. You have to ration to ensure that you are getting your money's worth out of the sweet substance, but you know that ultimately you could be going about this better.

But how? You can't go out into the woods and find a beehive now, could you? Not only is it difficult to find a honey hive, you would have to put up with the stings of thousands of bees who are all willing to die for the sake of their hive.

No, instead, you are going to have to go about this in another way.
You are going to have to keep your own bees.

Wait a minute, you think, how are you going to pull that off? Isn't keeping bees difficult? How do you make sure they are warm enough in the winter, and don't abandon the hive in the summer? Where do you even get bees to begin with?

With all these questions, it's easy to feel overwhelmed, but don't worry. If you are serious about starting your own beehive and beekeeping career, you have come to the right place.

I am going to show you everything you need to know to start – and maintain – your beekeeping hobby, and give you the secret to harvesting honey whenever you like.

This book is going to revolutionize the way you think about bees, and how you get your honey. By the time you reach the end, you are going to realize just how sweet this deal is.

Let's get started.

Chapter 1 – An Introduction to Beekeeping

If this is your first time beekeeping, it is likely that you are going to have a lot of questions. Odds are, you have seen beehives in farms or in fields before, though they are often too far away for you to see the bees themselves. Perhaps you have been fortunate enough to visit with friends or relatives who also keep bees, and you have been able to see the hives close up.

Unless you have engaged in beekeeping yourself in the past, now is the time to really get familiar with the hobby before you dive in.

People have been gathering honey since the dawn of time. With great skill (and a high pain tolerance) mankind has been known to have gathered honey for nearly 10,000 years. However, there is documented proof that Egyptians have domesticated and harvest bee honey for nearly 5000 years.

Though back in ancient times getting honey from bees was a more difficult task, modern day beekeeping needs to be well-researched, and you should really know what you are getting yourself in to before you begin. If bees are unhappy in their hive, or if they feel that they are in danger, they will swarm and find somewhere else to live, leaving you back at square one.

With this in mind, it is important that you know how to care for your bees before you bring them into the picture, so you can ensure they like the hive you prepare for them, and choose to stay there.

To start, it is important that you check into the local ordinances of where you live, and determine whether you need a permit for your hive or not. It's rare to need a permit to keep bees, however, it does happen, and you don't want to set up a hive, only to have it taken from you or to get fined later because you didn't know you needed a permit.

Next, understand that bees have their flight patterns, and that they are often follow the same routes when they are out gathering pollen. Their brains are wired to memorize maps of where the food is, and they will return to the same location day after day, gathering pollen and nectar to bring back to the hive (and turn into honey.)

With this in mind, place the hive in an area where the bees' flight pattern won't disturb your family, neighbors, or pets. Bees are very busy creatures, and they would rather not have to engage in any kind of activity with anyone other than the other members of the hive – so doing this is just as much a favor to them as it is to you.

Is beekeeping a lot of work?

The amount of work you put into beekeeping is going to vary. This is an incredibly seasonal hobby, meaning you are going to have virtually nothing to do in the winter, and you might find that it occupies all your free time in the spring. When it comes to bees, you are going to be in charge of the basic care, then let them handle the rest. Of course, you are going to get involved when it comes time to harvest the honey, but realize that bees are incredibly self-sufficient, and the less you are involved, the better.

When it comes to effective beekeeping, strategy is the best policy. Understand what a bee colony is and how it functions, understand the level of work you are going to have to put forth to ensure that your colony thrives, and understand how little (or how much) you are going to be involved to ensure that it is a success.

Learn how much honey you are allowed to take from the bees at a time, and remember to be generous with them when it comes to their honey. Only take what they have to spare and no more than that. There are a lot of rules that you need to follow, but trust me, once you get the hang of things, you are going to be set with your beekeeping hobby.

Time, diligence, and dedication are the three ingredients to beekeeping success.

Chapter 2 – Understanding the Colony

To ensure that you are going to give the bees everything they need, it is important that you understand what the colony is, and how it works. Though they may just seem like worker bees and a queen to us, there is actually an intricately designed system that is taking place, and each bee knows where they fit into that system.

Queen Bees

The queen bee is the only bee in the colony which lays fertilized eggs. Contrary to popular belief, she does not mate with the other bees throughout her lifetime, but rather, she mates early on in her life and stores the sperm within herself to continuously fertilize eggs throughout her life.

She is one of the most crucial parts to the colony, and without her, the colony will leave. Though it has been recorded that queen bees can live up to five years, a queen bee that is working hard to produce more eggs for the colony is more likely to live only two or three years before she passes on.

Colonies can grow to include thousands of bees – which is of little wonder when you learn that a queen bee can lay as many as two thousand eggs in just a single day.

Worker Bees

The queen bee lays the fertilized eggs in the colony, and when they hatch, they produce female worker bees. These bees are basically the backbone of the colony, converting the nectar that comes in with the other worker bees into honey, caring for the larvae, and making repairs around the colony as needed.

Though these worker bees can lay eggs, it is unusual for them to do so unless there is no queen present, or if the queen isn't producing as many eggs as she should be.

Drones

When there is no queen, or when the queen is unable to produce as many eggs as she ought, the worker bees will lay their unfertilized eggs, which then become the drones. The sole purpose of the drone is to fertilize the queen, and each male dies soon after mating.

It is not at all unusual for the worker bees in a colony to banish the male drones when the weather starts to cool, or when there is hardship in the colony. Though it is crucial that drones are in the picture from time to time, they really do place a small part in the activity of the colony as a whole, and are simply discarded by the rest of the bees when they have done their job.

Larvae

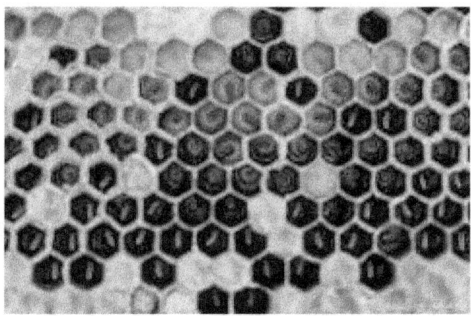

All bees begin as larvae, and undergo a scientific change which is known as metamorphosis. This is the process by which an egg is laid, then it is hatched into a larval. The larva then changes into a pupal, then they grow into their adult form.

A larval will shed its skin several times during this process before it finally emerges as an adult, and takes its place within the colony.

As I said, a colony is highly systematic, with each of the members doing what they are supposed to do and contributing to the good of the colony as a whole.

This is why it is important that you keep any interference with the colony to a minimum. Bees who are harassed tend to feel threatened, and they are willing to abandon a hive if it is for the overall good of the colony.

As the beekeeper, it is up to you to ensure that the main frame of the colony is taken care of, but as the beekeeper it is up to you to ensure that they are able to handle things for themselves.

Don't worry, I am going to show you how to care for your bees, and ensure both you – and they – are able to co-exist for a mutually beneficial relationship.

Chapter 3 – Getting Started the Right Way

The first step to successful beekeeping is to get what you need from the beginning. As I have already said, bees like to be left alone, and this means from the minute they are introduced to their new hive. You can plan on releasing the bees into the hive, then letting them get established on their own – something that you can imagine would be difficult if an intruder kept poking around.

So, to get started on the right track, you are going to have to ensure that you have all the supplies that you need before beginning.

This includes:

- **Hives – you can make these yourself using wood and other necessary supplies, or you can order kits that walk you through step by step** – For a first time beekeeper, I would recommend that you opt for a kit, as these are going to have everything you need (and in the proper size) for your bees.
-

You can find these kits virtually anywhere, including online department stores, Amazon, or private sellers (who are also online.)

- **Proper clothing** – if you want to harvest your honey without getting stung, you are going to need to have the proper clothing to do so. If you are harvesting honey the right way, you aren't going to have to worry about getting stung anyway, but this is added resistance that will greatly aid in the comfort you feel when you are harvesting.

- **Get a smoker and a hive tool** – a smoker and a hive tool are going to be the two things you use during the harvest. The smoker is going to put the bees to sleep or calm them down enough that you can get into the hive without too much trouble, and the hive tool is going to be used in the collection of the honey itself.

- **Once you have all the necessary supplies, source and order your bees.**

It is a good idea to go with a reputable bee seller for this, and understand that your bees are going to come in the mail. Though they are going to come packaged, this is still completely safe, as they are going to be enclosed in screen.

Keep in mind that your bees are going to be alive – meaning that they are going to have to survive the journey. Order from as nearby as you can to ensure there is minimal processing time – even if you were to overnight the bees from the other side of the country they are going to go through a lot of stress before they are delivered.

A common order for bees includes one queen bee and about three pounds of worker bees. The queen is going to be in her own box, so she is going to be easy for you to find and separate from the rest. Often, the bees also come with sugar water to keep them fed and hydrated throughout their journey.

Once the bees arrive, don't waste any time in getting them out into the hive where they belong. Open the side of the hive, giving you access to the inside, and pour the bees in. Once you open the box the bees came in, they are going to begin crawling out. Don't panic, simply pour them into the hive.

To make this process easier, begin with pouring syrup over the bars on the inside of the hive. This is not only going to give the bees something to do with themselves while they are put into the hive, but it is also going to keep them calm.

You may have to shake the box to ensure all the bees come out of the hive. Make sure the box is empty before you discard. There is going to be a special place for you to keep the queen bee – but leave her in her box. You don't want the bees to leave, and until they are used to the area, they are going to be in danger of doing so.

They will get used to living where the queen lives, so make sure she adopts the hive as her own. As the worker bees eat away the food that is blocking her entrance, they are going to get used to her as being one of the colony, and she is going to get used to the hive as being her own.

Keep in mind that the queen bee is raised separately from the rest of the colony, meaning they are not going to recognize her as one of their own on the outset. If you were to just place her in the hive outside the box with the rest of the bees, it is likely that they would kill her thinking of her as a foreigner – and thus a threat – to the colony.

By the time the worker bees do release her from her box, she is going to have released enough pheromones that the hive simply accepts her as their own, meaning they are going to take over with the mating and egg laying, and you just have to wait.

Though it's up to the worker bees to free the queen from her box, it is up to you to make sure that this happens. She can't live in the box indefinitely, and it is ideal for her to be out of there within the first few days.

Give the hive a few days to get settled, then come back and check on it. The queen should hopefully be released within the first 4 days, but if you find that she is still in her box, then loosen the hole a bit so the bees find it easier to get her out of the box.

After loosening the box, place the box back in its place and let it sit another 24 hours. Come back and check once again, repeating the process until she had been freed from her box.

You will find that the bees are eager to get to work, and they are very smart with how to go about doing that. It is highly unlikely you are going to have any problem getting the queen free, and in no time at all they are going to be making their rounds and scouting for flowers to begin making honey.

During the first couple weeks, make sure you provide your bees with plenty of sugar water. They are going to be so caught up in adjusting to the hive and getting the queen out of her box they aren't going to have time to go and gather food for themselves.

As long as the bees are cared for and don't have anyone harassing them, you are going to find they settle in quickly.

Chapter 4 – Harvesting the Honey

Once the time comes to harvest the honey, you will see that your hard work and dedication has paid off. However, before you go running out into your backyard with a pale and a scraper, it's important to take a moment to know what to expect – and to get ready.

First of all, you should get dressed in the appropriate attire. Wear gloves, a veiled mask and secure boots at a minimum, though I highly recommend that you also don be-proof overalls to complete the outfit, especially if you are a beginner. If you went through and purchased the list of things I recommended, you should already have one of these suits handy and ready to slip on.

The first thing you are going to do is smoke the bees.

Approach the beehive from behind, waving the smoker in front of and around you and the hive, ensuring that it is well fumed. Place the smoker at the entryway to the hive, driving the bees further in. This should be done in a fluid, steady motion – don't jerk around or move in fast, short movements.

Gentle strides, showing you are self-assured is best for both you and the bees. Once the bees have calmed down and retreated within the lower levels of the hive, pry open the top. This is going to take some effort as bees coat everything in a thick, sticky substance to ensure it stays together.

Be patient with it and don't force it, gently pry it away until the top peels back and you are able to get inside.

Next, you are going to remove the bees from the hive.

You can do this in a variety of ways, from scraping them off (gently!) to using a bee blower that will blow them off (again, gently!) to using your hands to brush them away. Just make sure that the panels are free of any bees and you are ready to pull them out.

You should have a clean bucket ready and waiting, then pull out the panels one at a time, pushing the honey into the bucket, and replacing the panels once you have cleaned them. Again, this should be done in a fluid, self-assured motion.

As you will notice, you are going to have a lot of honeycomb mixed in the honey – you can get rid of this using an extractor.

An extractor is somewhat like a salad spinner in that it spins quickly, causing the honey to fly to the edge of the drum and drip down into the center spigot.

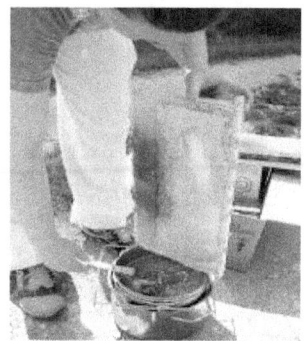

This is going to remove the combs from the honey, and ensure that you get the most for your hard work.

Bottle your honey next.

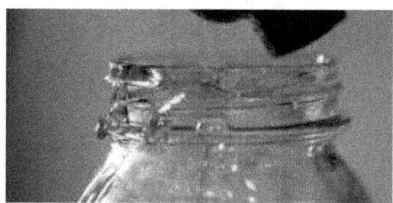

Make sure that you use honey that has been capped and matured by the bees – green honey is the newer batches, and though it might appear to be perfectly fine on the outside, it is actually prone to harboring yeast and bacteria and could cause your honey to spoil quickly – and once this happens, it is no longer safe for human consumption.

Now, many people question once they have harvested their honey whether they ought to then also pasteurize what they have gathered. This is a process that heats the honey to a high temperature, killing any potentially harmful bacteria or yeast which could be found in the honey.

While this is something that is entirely up to you to decide, it is important to note that doing this will also kill off some of the benefits that come with honey – so it's a decision you need to make for yourself.

Another thing to keep in mind when you are harvesting your honey is that it is always better to select quality honey over how much honey you are gathering, and it is important that you leave the bees with enough honey for themselves. There are many who keep bees who take more than they should, leaving the bees with a mix of sugar water or corn syrup – which the bees can't survive on.

Be fair with your portions, check the hives, and make sure everything is as it should be, and you are set!

Chapter 5 – Tips, Tricks, and Helpful Advice

That really is all you need to know to get started with your own beekeeping hobby – however, with as easy as it sounds, you must realize that it is going to take a lot of hard work for you to get your hive established and ready to go.

It's going to be worth it, and if you follow the methods that I have outlined here, you are going to have a thriving colony for years to come. However, even with the clearest of directions, there are still things you can do that will make this job that much easier, which is why I have included this final chapter of tips and tricks, plus some helpful advice that will make things run smoothly.

Try them out for yourself, and find what works for you – you just may be surprised.

Though bees are very much self-sufficient, when they are dealing with times of hardship, do your part and give them something to eat.

This can be syrup, sugar water, or candy – whatever works for you – just keep in mind that you should do this when the bees are having a hard time gathering enough nectar for themselves, not because you have taken too much of their honey.

 As I said, working together with this relationship with your bees is the one thing that is going to ensure both of you are happy in the long run.

Feed your bees the syrups through a plastic bag to ensure they are able to reach the food, but they aren't going to fall in and drown.

To do this, you are going to fill a plastic Ziploc (or similar) bag with sugar water, syrup, or whatever it is you have chosen to use to feed your bees, then you are going to place it on the top rack in their hive. You are then going to take a sharp pair of scissors or a razor, and you are going to create slits along the top portion of the bag.

This will allow the bees to stand on top of the feeder without being in danger of falling in.

Don't forget the water.

In addition to their honey and the gathering that they do, the bees you keep are going to need water. Now, watering bees is rather tricky, as they are very picky about where they drink.

Try setting out a dog dish next to the hive and make sure to keep water in it at all times. Make sure any pets you have have another source of water so they aren't forced to go near the hive when they are in need of a drink.

If you have a problem with your bees swarming, trying replacing the queen every year.

Though many bees will be happy to reside in a certain hive for years, there are times when you may have a difficult time keeping your bees from leaving. This is because some bees enjoy swarming when the hive has grown strong, and they are always on the move for a new location with more food.

If you find that this keeps happening with your bees, simply replace the queen with a new, young queen every year. A queen that is less than a year old is going to provide plenty of eggs for the colony, but she is far less likely to swarm with the colony at her young age.

To introduce a young queen, you are going to follow the same steps that you did when you bought the bees in the first place. Be patient with them, and you are going to get the results that you want.

Put a fence around your hives.

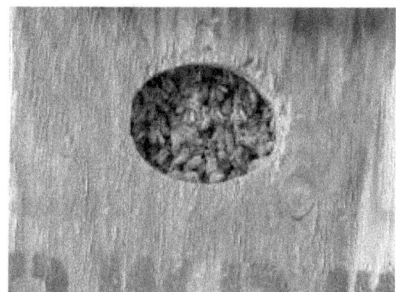

Ideally, this will be a six foot tall wooden fence or shrubbery, or something of that sort. This is going to force the bees to fly over everyone rather than through them, and it is going to make it less likely that anyone is going to bother your bees.

At the same time, it is going to give them protection from the wind, making it easier for them to come and go.

Try out all of these tricks regardless of how old your hive is, and you are going to have a successful colony (and a steady supply of honey) for years to come!

Conclusion

There you have it, everything you need to know to create your own bee colony, and how you can ensure they are going to be cared for and give you as much honey as you could ever use for years to come. I hope this book inspires you, whether you are a first time keeper or if you are adding on or starting up after a period of not expanding.

You are going to quickly learn that beekeeping is a hobby that is highly addictive, and regardless of how long you have been doing it, you will want to continue doing it and expanding your bees. The more bees you have, the stronger your colony is going to be, and the more honey you are going to get in the long run.

There really is no way you can ever complete this hobby, as you are going to start a cycle of caring for your bees and harvesting the honey – this will last as long as you want it to, and it will give you an organic, constant supply of the best honey on the planet.

You'll get to know your bees, and you will learn just how to care for them to ensure longevity and optimal care. If you pay attention to what you are doing, you really can't do it wrong, and this book is going to ensure that you get the help you need on the outset to guarantee the best results going forward.

It doesn't matter if this is your first colony or your hundredth, you are going to learn something new and useful here, giving you the boost you need to have a thriving colony for years to come. You are going to fall in love with not only the quality of the honey, but the process with which you get it.

Do what is right by your health, for your family, and for the planet, and you are going to see that organic, homegrown honey is truly the way to go. Your bees will thank you, and you will help preserve a species that is starting to face hardship.

Don't be afraid to put in the work, and you are going to be rewarded for your efforts, guaranteed. So gather the supplies you need, plan the work, then put your plan into action, and within months you are going to have that constant supply of honey you have been dreaming of.

Your health will increase, and your friends and family will thank you – trust me, it's not going to take long before you realize that this is all more than worth it. Good luck, and happy beekeeping.

www.ingramcontent.com/pod-product-compliance
Lightning Source LLC
Chambersburg PA
CBHW071731170526
45165CB00005B/2245

ISBN 9781548024253